SPACE DISCOVERIES

MISSION: MARS

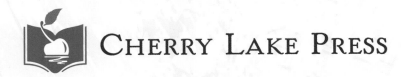

Published in the United States of America by Cherry Lake Publishing
Ann Arbor, Michigan
www.cherrylakepublishing.com

Reading Adviser: Beth Walker Gambro, MS, Ed., Reading Consultant, Yorkville, IL
Book Designer: Book Buddy Media
Photo Credits: Cover: ©nasa.gov; page 1: ©24K-Production / Getty Images; page 5: ©NASA/JPL-Caltech / nasa.gov; page 7: ©JUAN GAERTNER/SCIENCE PHOTO LIBRARY / Getty Images; page 9: ©NASA/JPL-Caltech/MSSS / nasa.gov; page 10: ©nasa.gov; page 13: ©MARK GARLICK/SCIENCE PHOTO LIBRARY / Getty Images; page 15: ©nasa.gov; page 16: ©nasa.gov; page 19: ©NASA/JPL-Caltech / nasa.gov; page 21: ©NASA/JPL-Caltech / nasa.gov; page 23: ©NASA/JPL-Caltech / nasa.gov; page 25: ©STS-119 Shuttle Crew/NASA / nasa.gov; page 26: ©NASA/JPL-Caltech / nasa.gov; page 27: ©NASA/JPL-Caltech/FFI / nasa.gov; page 29: ©nasa.gov; page 30: ©NASA/JPL-Caltech / nasa.gov

Copyright ©2022 by Cherry Lake Publishing Group
All rights reserved. No part of this book may be reproduced or utilized in any form or by any means without written permission from the publisher.

Cherry Lake Press is an imprint of Cherry Lake Publishing Group.

Library of Congress Cataloging-in-Publication Data has been filed and is available at catalog.loc.gov

Cherry Lake Publishing would like to acknowledge the work of the Partnership for 21st Century Learning, a Network of Battelle for Kids.
Please visit *http://www.battelleforkids.org/networks/p21* for more information.

Printed in the United States of America
Corporate Graphics

ABOUT THE AUTHOR

Mari Bolte is a children's book author and editor. Streaming sci-fi on TV is more her speed but tracking our planet's progress across the sky is still exciting! She lives in Minnesota with her husband, daughter, and a house full of (non-Martian) pets.

TABLE OF CONTENTS

CHAPTER 1
Destination: Mars............................... 4

CHAPTER 2
Traveling Through Space 8

CHAPTER 3
X Marks the Spot 14

CHAPTER 4
Exploring the Planet 18

CHAPTER 5
The Future 24

ACTIVITY.. 30
FIND OUT MORE 31
GLOSSARY 32
INDEX... 32

CHAPTER 1

Destination: Mars

Before 1000 BCE, ancient people saw a fiery ball among the stars. All they knew was that it moved around in the sky. The first **astronomers** observed the planet's path. Egyptians knew that Mercury, Venus, Jupiter, and Saturn moved in a similar way. Greeks called the red planet Ares, after their god of war. Romans called the planet Mars, after their own war god.

Mariner 4 spent about 3 years circling Mars. It studied the planet's solar wind.

Modern humans got the first view of Mars in 1965. The National Aeronautics and Space Administration's (NASA) *Mariner 4* sent back photos as it passed the planet. The Viking project made history in 1975 when it landed the first spacecraft on the Red Planet. *Viking 1* also sent back the first photos taken from Mars's surface.

In 1997, the rover *Pathfinder* rolled its way across Mars. The helicopter *Ingenuity* made its first flight in 2021. Will humans setting foot on Mars be the next huge accomplishment?

Technological Advances

To fly beyond Earth's atmosphere, spacecraft use multiple stages. Stages are sections of rockets that fire in a specific order. When a stage is used up, it drops off the spacecraft. Then it falls back to Earth. Until the 1980s, spent stages would burn up in Earth's atmosphere. But then, scientists began designing stages that could be recovered and reused. Some believe reusing equipment makes the flight cheaper. Some companies are even building rocket stages that will return to Earth and land themselves.

With every trip to Mars, scientists have learned a little more about how to make the next one a success. Many early missions were lost on the way or failed to land successfully. About half of them ended in failure. Now, though, the success rate is much higher. All of NASA's spacecraft since 1999 have arrived safely. They have also kept working long after their missions ended.

Sharing information about space travel has helped scientists around the world succeed. The International Space Station (ISS) coordinates flight crews and launch vehicles. They share engineering, training, and scientific research with each other. They use the same communications network and testing labs. Astronauts from around the globe live and work together aboard the ISS. The things they learn can be applied to future missions to Mars.

Mars is the only planet in our solar system with carbon dioxide snowfall.

Extreme Skiing on Mars

The average temperature on Mars is −80 degrees Fahrenheit (−60 degrees Celsius). This means that it is cold enough for snow. But **carbon dioxide** snow on Mars is different from the frozen water crystal snow on Earth. Snowflakes on Mars are either microscopic or tiny ice **particles**. Martian snow does not have much water. It would not pack well, so there would be no snowball fights on Mars. But it would be great for snow sports! Fresh carbon dioxide snow might be more slippery, which would make up for the flatter Martian surface. Getting air for freestyle jumps would be a lot easier! To add another extreme element, carbon dioxide snow almost always falls at night. Night vision goggles would add an extra level of extreme.

CHAPTER 2

Traveling Through Space

The idea of sending people to Mars has been a recent goal. But we have learned a lot about the planet already. The first known meteorite from Mars hit Earth in 1815. Then another fell in 1865. But our biggest payload from Mars came on June 28, 1911. Forty pieces of a huge **meteoroid** fell from the sky that day near the village of El-Nakhla, Egypt. At the time, scientists did not know they were Mars rocks. Some even believed rocks from Mars would melt if they left the planet's atmosphere. It would be another 50 to 60 years before any real scientific study was actually done to these Nakhla rocks.

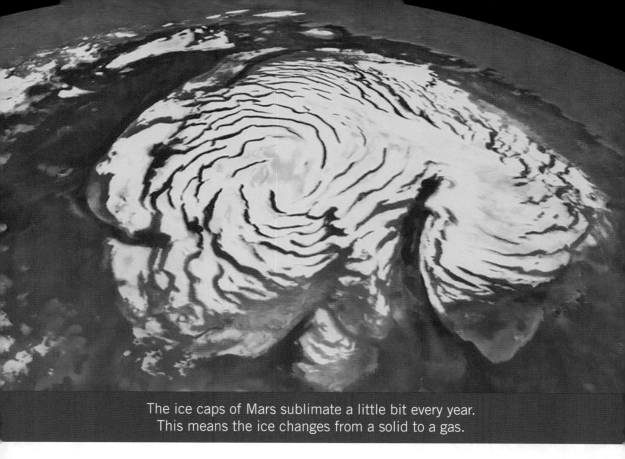

The ice caps of Mars sublimate a little bit every year. This means the ice changes from a solid to a gas.

Then the two *Viking* spacecraft sent back atmospheric data in 1976. Finally, scientists knew that rocks would not melt if taken off of Mars. In 1983, NASA compared another Martian meteorite to the *Viking* data. They measured gases found in the meteorite to gases found on the planet. This proved the meteorite really was from Mars. Then they used that data to look at the Nakhla pieces again. Similarities made it clear that the Nakhla meteorite was also from Mars. Studying the meteorites helped scientists understand the Red Planet's **geology**.

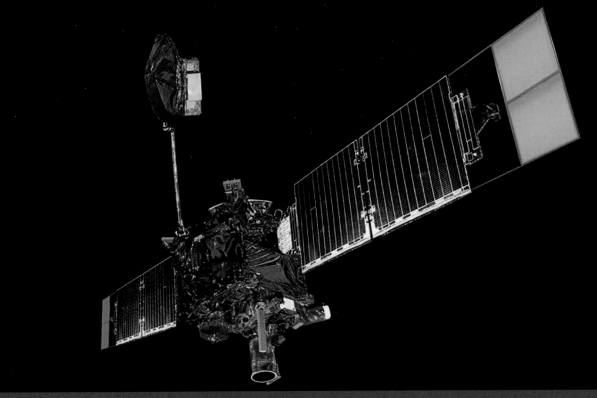

The *Mars Global Surveyor* launched on November 7, 1996. It reached Mars on September 11, 1997 on a quest to find signs of water in the past. The surveyor lost contact with Earth in November 2006.

Finding water on Mars has always been a major goal for scientists. In 1923, they saw canals and **polar ice caps** in a telescope. Without photo evidence, though, not everyone believed they were ice caps. Even when those photos were finally published in the 1960s, many people were doubtful. Some said the "ice caps" at the poles were made of frozen carbon dioxide. *Viking 2* confirmed that the ice caps were made of frozen water.

Nothing is Wasted

To **colonize** Mars, people would need water to drink, stay clean, and grow food. Scientists are still not sure what would happen if the ice caps melted. They also do not know if there is actually usable water on Mars. So explorers would need to bring or create their own water source. Scientists are still working on the problem of how to store and clean water on Mars. Right now, about 90 percent of water-based liquid on the ISS is recycled. This includes urine and sweat. "When it comes to our urine on the ISS, today's coffee is tomorrow's coffee!" one astronaut said.

In 2000, the *Mars Global Surveyor* sent photos of dried-up lake beds. It worked for 10 years, spending most of its time in polar orbit around the planet. It took daily images to show how the planet changed over time. The photos also showed places where water had flowed. The rover *Opportunity* studied rocks on Mars. Scientists determined the rocks had been around water in the past.

Cloudy with a Chance of Dust

When you see a dark cloud in the sky, you might turn on a local weather report or check a weather app to see if a storm is coming. Meteorologists use tools that measure things like atmospheric pressure, temperature, and wind speed. They look at that data. Then they make educated guesses to create weather **forecasts**.

In 2018, scientists found a large saltwater lake on Mars. It was hidden under the ice at the south pole. Two years later, the European Space Agency's (ESA) **orbiter**, *Mars Express*, found it, too. It also found three more. The orbiter used **radar** to send out radio waves. The waves bounced off the planet's surface and subsurface. The way the waves returned to the spacecraft told scientists what the planet's surface was like. For example, it told them what areas were made of rock, ice, or water.

The dust storms on Mars are so huge that some can be seen through telescopes on Earth.

Right now, *Perseverance* is collecting similar information. It is working with the *InSight* lander and the rover *Curiosity*. Together, they are building a meteorological network on Mars. It will not only tell future astronauts the weather. It will also give them information about the dust particles in the air. Martian dust storms are powerful enough to shut down solar panels and reduce the amount of solar **radiation** that reaches the surface.

CHAPTER 3

X Marks the Spot

Our future home is being researched and observed every day. In 1971, *Mariner 9* was the first spacecraft to orbit Mars. It arrived during a global dust storm. But once the storm cleared, it showed us volcanoes and canyons. It also showed us the polar ice caps that gave scientists hope that the planet had water.

The *Viking* orbiters got there in 1976. They showed us the many landforms that make up the planet. They mapped weather patterns. They took photos of Mars's moons, Deimos and Phobos. And they communicated with the landers that traveled there with them.

Mariner 9 mapped 85 percent of the planet while in orbit. It sent more than 7,000 images back to Earth.

There was a 20-year gap in successful missions. Then, in 1997, the *Mars Global Surveyor* entered Mars's orbit. It studied the entire planet's surface. It gave scientists the first 3D images of the polar caps. It sent back daily images that help scientists better understand the planet's weather patterns. The spacecraft helped the *Spirit* and *Opportunity* rovers communicate with Earth until it stopped working in 2006.

HiRISE can capture thousands of images of Mars in just a few seconds.

Launched in 2001, *Mars Odyssey* is NASA's longest-lasting Mars spacecraft. Its main mission ended in 2004, but it is still operating today. It made the first global map of the chemicals and minerals that make up the planet's surface. It also collected and sent information from the *Mars Phoenix Lander* and the rovers *Spirit*, *Opportunity*, and *Curiosity*.

The *Mars Reconnaissance Orbiter* has been orbiting since 2006. It carries a powerful camera called the High Resolution Imaging Science Experiment, or HiRISE. HiRISE has sent back images that help scientists understand the history of

Jupiter

Jupiter is the fifth planet in our solar system. It is also the largest. Its mass is more than double all the other planets combined. If Earth was the size of a marble, Jupiter would be the size of a basketball. Jupiter has the shortest days and the longest years of any planet in the solar system. One day is only about 10 hours long. But a year on Jupiter is nearly 12 Earth years. Jupiter is made of gas. There is no solid ground, just layers of swirling gases. It has more than 75 moons and very faint rings. Its Great Red Spot is a giant storm. The spot is two times larger than Earth. It has been storming for hundreds of years.

water on Mars. The orbiter also serves as a communication hub between Earth and Mars. It takes information from rovers and other obiters. Then it sends it back to the Deep Space Network, which are huge antennas on Earth.

The orbiters have taught us about the surface of Mars. They have recorded weather patterns and looked for water. Because of them, we know more about Mars's atmosphere and radiation levels. And they will be there to help human explorers the second they set foot on the planet's surface.

CHAPTER 4

Exploring the Planet

Orbiters and rovers have used their tools to look at the rocks on Mars. *Perseverance* is there now, collecting rock and soil samples. It is looking for areas that may have been in contact with **microbes**.

Rocks and minerals record what the environment was like when they were crystallized. Some types are formed under extreme heat or pressure. Some can store chemical traces left behind by life billions of years ago. Others can only be formed when water is present. These types of rocks are called sedimentary rocks. The bottom of Valles Marineris, the huge canyon that stretches across Mars, is covered with beds of layered sediment. On Earth, these rocks would have formed at the bottom of lakes and shallow seas.

Valles Marineris is 2,485 miles (4,000 km) long. That is long enough to stretch across the continental United States.

Math on Mars

Perseverance's sample tubes are less than 6 inches (15.2 centimeters) long and weigh less than 2 ounces (57 grams) each. One tube can hold a sample about 2 inches (5 cm) long. That is about the size of a piece of chalk. Each sample weighs about half an ounce (15 g). The tubes are coated to protect the samples against the Sun's rays. Each is labeled with a special serial number. The number will help researchers identify each sample once they are back on Earth.

Perseverance is one of the most complex robots ever built. It is easy to think of *Perseverance* as one roving robot. But it's really more like multiple robots working together. Its Sample Caching System alone is three separate robots. They are made up of more than 3,000 parts.

Perseverance brought 43 sample tubes to Mars. They are similar in size to test tubes. Thirty-eight of the tubes will collect fresh samples. The remaining five tubes are "witness" tubes. They were preloaded with sample material that will collect tiny contaminants in the air. The tubes will be opened near the

Perseverance uses an auto-navigation system that lets it think while driving. This could allow it to hit a speed as fast as 393 feet (120 meters) an hour.

sample testing site to "witness" the environment conditions. If there are any contaminants from Earth present, the tubes will collect them, essentially recording their presence. Hopefully, they will help scientists find out in advance if any gases or chemicals were released from the rover or rode along from Earth. Then there will be no confusion about what originated on Mars.

Other Exploration

Rocks and dirt aren't the only kinds of samples scientists take. Inventors are working on a robot that can take blood from people. Blood samples are a simple part of medical testing. But it can be difficult to get a good sample if the patient's veins are hard to reach. Repeated needle sticks can cause soreness and bruising, nerve damage, or even infection. And accidental needle sticks can be dangerous for healthcare workers. A robot would not face that risk. The robots could also be used for inserting needles used to give fluids or medication through a vein.

Perseverance's test tubes are part of the Mars Sample Return mission. Eventually, NASA and the ESA will send a "fetch" rover. This rover will pick up the samples *Perseverance* took. Then it will bring them to a small rocket called the *Mars Ascent Vehicle (MAV)*. The *MAV* will have a special container for the samples. The *MAV* will shoot a rocket loaded with the samples into orbit. A **satellite** known as *Earth Return Orbiter* will

The *MAV* has size limitations. It must be no taller than 9.2 feet (2.8m) tall and no wider than 1.9 feet (75 cm). It also must weigh less than 881 pounds (400 kilograms).

catch the container before heading back to Earth. The Mars Sample Return mission will cost around $7 billion to complete.

The fetch rover is due to launch in 2026. By 2029, the samples will be fired into space. Then they should get to Earth by 2031. Finally, scientists will have real samples to handle in person.

CHAPTER 5

The Future

Life on the ISS and the information orbiters and rovers send back to Earth has given us a lot to learn. A future on the Red Planet is drawing near. Mars 2020, the mission that includes *Perseverance* and *Ingenuity*, uses the most advanced technology available so far. The mission's spacecraft had **sensors** on its landing equipment. Sensors are devices that detect or measure changes. What they learn can help scientists figure out how to improve their design when the time comes to land heavy equipment needed for living on Mars. At 2,260 pounds (1,025 kg), *Perseverance* itself is the largest rover to ever land on Mars's surface. It is the size of a small car.

The United States, Russia, Canada, Japan, and Europe own and operate the ISS. Each country is responsible for the parts of the ISS that they provide. For example, European law applies in the European laboratory. The countries work together to share equipment in a fair way.

Sharing Space

The US space program, NASA, was created in 1958. Since then, NASA has made more than 3,000 agreements with 125 nations. About half of those agreements were made with eight other countries. Those partners are France, Germany, the ESA, Japan, the United Kingdom, Canada, Italy, and Russia. The ESA is part of the mission to retrieve the samples taken by Perseverance. NASA uses Russian equipment to send astronauts to the ISS. There is no space exploration program as politically complex as the ISS. Astronauts from 19 different countries have visited the ISS. Will the world's powers be able to work together on Mars in the same way? Only time will tell.

Scientists use RIMFAX's radar to figure out the different kinds of rock, ice, or other layers that are underground.

Perseverance is also carrying a radar tool. The Radar Imager for Mars' Subsurface Experiment (RIMFAX) uses radar waves to look at the ground. It doesn't just look at the ground right under the rover's tires, though. It uses ground-penetrating radar that can "see" more than 30 feet (10 m) below the surface. It's like x-ray vision, but for under the ground! The radar shows the layers that make up the planet and what they are made of. RIMFAX can also find ice, water, or salty brine. What it finds could be used by future colonists to the planet.

RIMFAX is the first radar tool that NASA sent to the surface of Mars.

Orbiters use radar to give scientists a broad picture of what's under Mars's surface. They can penetrate deeper underground, but don't get as detailed a picture. Because RIMFAX is on a rover, it picks up small changes in the **terrain**. As *Perseverance* moves, RIMFAX collects data every 4 inches (10 cm).

Unread Messages

On Earth, many of the things we do rely on the hundreds of satellites that orbit the planet. Emergency workers use navigation to find where they need to go. News updates are sent out as soon as an event happens. Satellites allow stores to communicate with banks to collect money. Things that happen almost instantly on Earth would not be possible on Mars. There would be no way to stream TV shows or movies. Debit and credit cards would be useless. And forget updating social media! Living on Mars would mean using the latest technology for space exploration—without having what we consider the latest technology every day. Eventually, similar services could be built on Mars. But it will take time to get them up and running. Until then, your messages would have to stay unread until you returned to Earth.

Traveling to Mars comes with a lot of questions and a lot of problems to solve. How will we get there safely? How will we breathe? What about food and water? Will we be able to withstand the Sun's radiation and the Red Planet's dust storms? Scientists at NASA and around the world are working hard to find the answers. More than a billion dollars and years of rover exploration keep bringing new discoveries. NASA's Artemis Program plans to land the first woman and the first person of color on the Moon. No doubt Mars is next on the list!

NASA's Space Launch System will be the most powerful launch vehicle ever built. It can launch four-member crews deep into space.

Activity: Sample Retrieval

Once *Perseverance* sets down its sample canisters, it cannot pick them back up again. So how will the fetch rover know where to look? Scientists are using landmarks on the ground and measurements to keep track of each canister. Practice recording information using similar techniques here on Earth.

WHAT YOU'LL NEED:

- small, colorful objects, such as painted rocks or toys
- paper and pencil
- measuring tape
- compass

1. Write a description of each object. What are its dimensions? What color is it? What shape is it? Are there any additional details that would help someone identify the object just from your description?

2. Hide the first object. Then study your surroundings. Find landmarks, such as trees, tables, desks, or plants. Measure the distance between each landmark and the object. Use the compass to determine direction. Give locations from multiple directions.

3. Once your items are logged, hidden, and their locations are recorded, switch information with another person or group. Can you find each other's items?

Example: Item is a blue-and-white flat rock, 2 × 3 inches (5 × 7.6 cm), dark underneath. It is located 10 feet (3 m) west of the pine tree, 2 feet (0.6 m) south of the sunflower patch, 16 feet (4.9 m) northeast of the red bench. Look near the edge of the gray gravel.

Find Out More

BOOKS

Bearce, Stephanie. *This or That Questions About Space and Beyond: You Decide!* North Mankato, MN: Capstone Press, 2021.

Downs, Mike. *Imagining Space.* Vero Beach, FL: Roarke Educational Media, 2021.

Kenny, Karen Latchana. *Breakthroughs in Planet and Comet Research.* Minneapolis, MN: Lerner Publishing Group, 2019.

Loh-Hagan, Virginia. *Mars Colonization.* Ann Arbor, MI: Cherry Lake Publishing, 2020.

WEBSITES

NASA: 10 Amazing Space Discoveries by the World's Largest Flying Observatory
https://nasa.tumblr.com/post/618561393615110145/10-amazing-space-discoveries-by-the-worlds
NASA's Stratospheric Observatory for Infrared Astronomy (SOFIA) is a flying observatory that carries a telescope to study the solar system from anywhere in our world.

NASA: Artemis Program
https://www.nasa.gov/specials/artemis
Learn about NASA's Artemis program, which plans to land the first woman and the first person of color on the Moon.

National Geographic: Planet Mars, Explained
https://www.nationalgeographic.com/science/article/mars-1
Full of facts about the mysteries of Mars and the discoveries made by science.

Science Alert: What Is Mars?
https://www.sciencealert.com/mars
Get the details on the Red Planet and discover what a day on the Red Planet is like.

GLOSSARY

astronomers (uh-STRON-uh-muhrs) people who study astronomy

carbon dioxide (KAR-buhn die-OX-ide) a heavy colorless gas; humans inhale oxygen and exhale carbon dioxide

colonize (KOL-uh-nyz) to send a group of settlers to a new place

forecasts (FOR-kasstz) predictions of future weather or other events

geology (gee-AH-luh-gee) the makeup and physical features of an area

meteoroid (MEE-tee-ohr-oyd) a rocky body traveling through space; they become meteors once they enter a planet's atmosphere, and meteorites once they hit land

microbes (MY-krowbs) single-celled microorganisms that are too small to be seen by the naked eye; microbes include bacteria, algae, and amoeba

orbiter (or-BIT-uhr) a spacecraft designed to circle a planet

particles (PAR-tuh-kuhlz) very small objects

polar ice caps (PO-luhr EYES-kaps) dome-shaped masses of glacier ice

radar (RAY-dahr) a system for detecting objects by sending out pulses or waves; the waves are reflected off the object and back to the source

radiation (ray-dee-AY-shuhn) a form of energy that travels through space

satellite (SAT-uh-lite) an artificial body placed in orbit

sensors (SEN-sohrz) devices that detect or measure changes

terrain (tuh-RAYN) the physical features of an area

INDEX

Artemis Program, 28

European Space Agency (ESA), 12, 22, 25

ice, 7, 10, 11, 12, 14, 26, 32
Ingenuity, 5, 24
International Space Station (ISS), 6, 11, 24, 25

landers
    InSight, 13
    Mars Phoenix Lander, 16

missions
    Mars Sample Return, 22, 23
    Viking, 5, 9, 10, 14

Nakhla meteorite, 8, 9

orbiters
    Mariner 4, 5
    Mariner 9, 14
    Mars Express, 12
    Mars Global Surveyor, 11, 15
    Mars Odyssey, 16
    Mars Reconnaissance Orbiter, 16

rovers
    Curiosity, 13, 16
    Opportunity, 11, 15, 16
    Pathfinder, 5
    Perseverance, 13, 18, 20, 22, 24, 25, 26, 27, 30
    Spirit, 15, 16